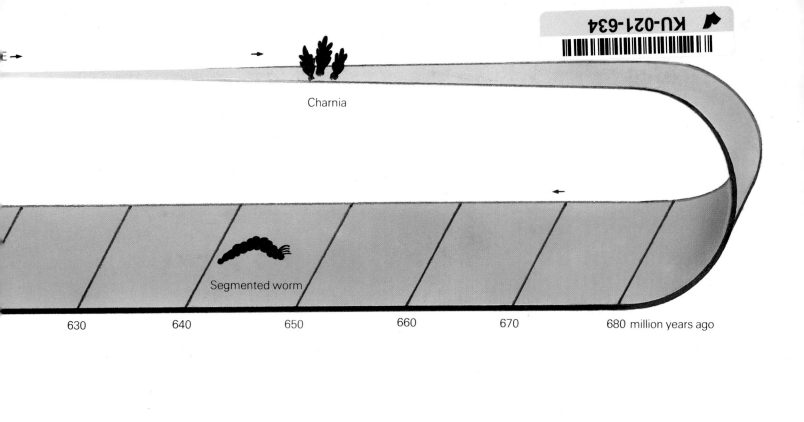

Charnia

Segmented worm

630   640   650   660   670   680 million years ago

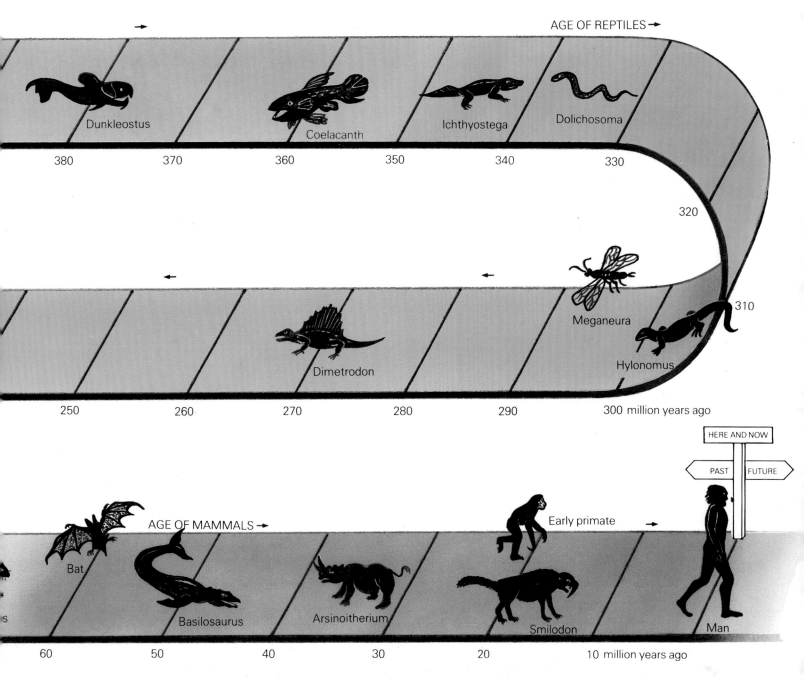

AGE OF REPTILES ➔

Dunkleostus

Coelacanth

Ichthyostega

Dolichosoma

380   370   360   350   340   330

320

Meganeura

310

Dimetrodon

Hylonomus

250   260   270   280   290   300 million years ago

HERE AND NOW

PAST   FUTURE

Early primate

AGE OF MAMMALS ➔

Bat

Basilosaurus

Arsinoitherium

Smilodon

Man

60   50   40   30   20   10 million years ago

Designed by Brigitte Willgoss
Edited by Debbie Lines

ISBN 0 86112 668 8

Published by Brimax Books Ltd., Newmarket, England 1989.
This book was previously entitled Animals from the Dawn of Time.
Second printing 1990.
Printed in Portugal by EDIÇÕES ASA—DIVISÃO GRÁFICA

4000 MILLION YEARS AGO

Written by Stephen Attmore
Illustrated by David A. Hardy

Contents

Brimax Books · Newmarket · England

When was the Dawn of Time?

The story of life on Earth goes back to the very dawn of time
– over 3,500 million years ago.

It began with the microscopic life-forms that came into
being in the orange-coloured 'soup' that made up the seas
that covered the planet.

From that starting point many different forms of life
developed. Some, like the dinosaurs have disappeared along
the way. A few have survived and exist today much as they
were thousands or even millions of years ago; others have
undergone changes as part of the process of evolution.
Together they form the wide range of animal species alive
today.

This book traces the development of animal life from the
very beginning, showing which forms of life dominated the
world at different stages. Within the book, those animals
featured in boxes are surviving examples or possible
descendants of prehistoric species. Those animals featured in
bold type are illustrated.

How to say the names of the prehistoric creatures

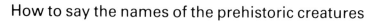

algae	al-gee	hyracotherium	hi-rak-oh-therr-ee-um
archaeopteryx	ark-ee-op-ter-iks	icarosaurus	ik-arrow-sor-us
australopithecus	ostra-loh-pith-eek-us	ichthyornis	ik-thi-ornis
brachiosaurus	brak-ee-oh-sor-us	ichthyostega	ik-thi-oh-steeg-ah
charnia	char-nee-ah	mastodonsaurus	mass-toe-don-sor-us
cladoselache	clad-oh-se-lak-ee	meganeura	meg-ah-nure-ah
coelacanth	see-la-kanth	megatherium	meg-a-therr-ee-um
coelophysis	see-lo-fy-sis	megazostrodon	meg-ah-zos-troh-don
coelurus	see-loor-us	megistotherium	meg-isto-therr-ee-um
compsognathus	komp-so-na-thus	paramys	pah-rah-mis
cynognathus	sino-na-thus	podopteryx	pod-op-ter-iks
deinonychus	die-non-ike-us	proganochelys	pro-gan-oh-kel-iss
deinosuchus	die-no-sook-us	pteranodon	ter-a-no-don
diadectes	die-ah-dek-tees	pteraspis	ter-asp-iss
diatryma	die-ah-try-ma	quetzalcoatlus	kwet-zal-kote-lus
dinichthys	din-ik-thees	ramapithecus	ram-ah-pith-eek-us
diplocaulus	dip-loh-cor-lus	rhamphorhynchus	ram-for-in-kus
dolichosoma	dol-ik-oh-som-ah	saltoposuchus	salt-oh-poss-ook-us
dunkleostus	dunk-lee-oss-tis	scutosaurus	skew-toe-sor-us
dryopithecus	dry-oh-pith-eek-us	stegoceras	steg-oss-er-as
echidna	e-kid-na	synthetoceras	sin-thet-oss-er-as
ellopos	ee-lop-oss	tanystropheus	tan-ee-strofe-ee-us
eryops	erri-ops	triceratops	try-ser-a-tops
eusthenopteron	yous-the-nop-terr-on	tuatara	twa-tar-a
hesperornis	hess-per-or-niss	tyrannosaurus	tie-ran-oh-sor-us
hylonomus	hi-lon-o-mus	uintatherium	win-tah-therr-ee-um

When did prehistoric animals first appear?

Tiny living things appeared on Earth 3500 million years ago. Remains of the earliest known creatures are found in rocks dating back nearly 600 million years. The first animals, which were simple fish, evolved about 530 million years ago.

About 200 million years later came the invasion of the land by the sea creatures that developed into amphibians. The age of the reptiles followed with the mighty dinosaurs pre-eminent for 160 million years. When they died out they gave way to birds and mammals. From among the mammals about 10 million years ago a small ape began to walk upright evolving eventually into the species homo sapiens – the human beings of today.

The Earth 600 million years ago

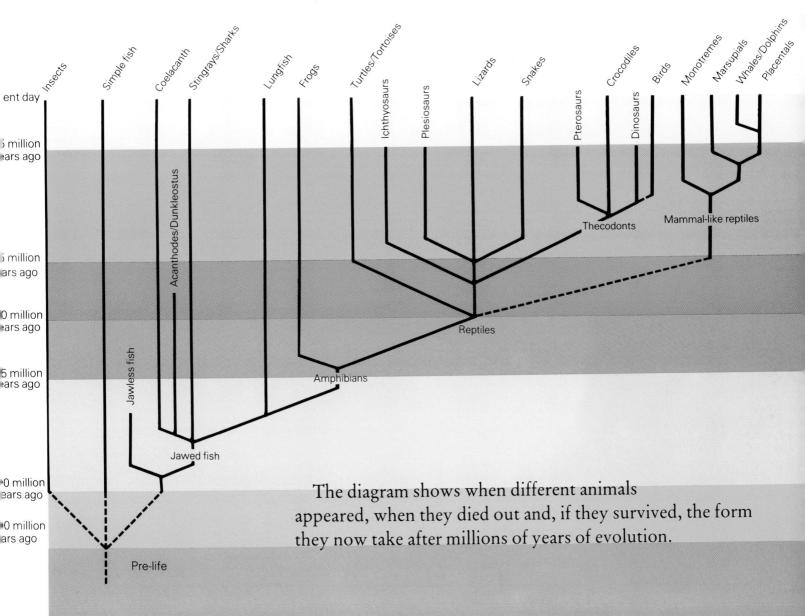

The diagram shows when different animals appeared, when they died out and, if they survived, the form they now take after millions of years of evolution.

The first forms of life

3,500–550 million years ago

4,500 million years ago, Earth was a very different planet to the one we live on today. The outer surface of the planet was gradually cooling as the crust slowly hardened. Thunderstorms shook the air and many volcanoes erupted. Thousands of years of rain created warm shallow oceans. The sun poured lethal radiation over the planet. There was no oxygen in the air.

Over millions of years the seas became an orange-coloured 'soup', full of chemicals that had been washed off the land. In this soup, small living things began to develop. These were single cells, called **bacteria** – the smallest and simplest forms of life. Some of these developed a green chemical, chlorophyll, to make their food. They used this and sunlight to convert carbon dioxide in the air into food. They also released oxygen into the air which is vital for all living things.

Over many millions of years the amount of oxygen increased, eventually forming a screen against the sun's rays. This is called ozone and it is what makes the sky look blue. Conditions were now right for more advanced life to begin.

Simple plants such as **blue-green algae** (1) were among the first forms to appear. The oldest known fossil is of algae, 3,400 million years old. Algae like this still survives today. You can see it on the top of ponds, making the surface look green. After about 2,500 million years, cells began to group themselves together to make very simple plants and animals. Here the cells eventually became heart, brain and muscles. **Jellyfish** (2) was an early animal. Fossilised remains in rocks have been dated as 600 million years old.

By about 570 million years ago many invertebrates (animals without backbones) were active on the sea-bed. There were worms, molluscs, sea-snails and arthropods. We know about these creatures from tracks, tunnels and prints of their bodies left in some rocks.

Spriggina (1) was a segmented worm. Its body was 5 cm (2 inches) long. The **trilobite** (2) was a large arthropod. Each segment on a trilobite's body had a pair of limbs which it used for walking, swimming, breathing and handling food. Trilobites died out about 250 million years ago. **Dentalium** (3) was a mollusc with a fleshy 'coat' and a hard shell. **Charnia** (4) was a soft coral fixed to the sea-bed, 40 cm (1⅓ ft) tall. **Sponges** (5) fixed themselves to a rock and filtered water for food.

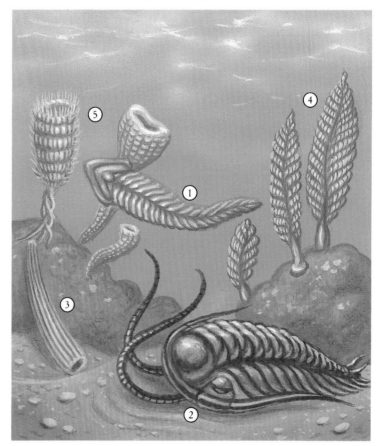

The fossilised remains of **sponges** have been found in rocks as much as 395 million years old. They have hardly changed since then. A living sponge is a sort of fleshy bag held up by a skeleton. For many years scientists thought sponges were plants. Then they realised that a sponge breathes and feeds. It draws in water through lots of little holes before removing tiny particles of food.

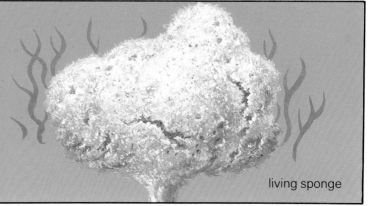

living sponge

From the sea to the land

550–345 million years ago

Let us go back to about 550 million years ago. From fossils found in a mudstone cliff in the Rocky Mountains in Canada it is possible to picture a scene under water. Here were some of the first animals which swam in the seas. All of the creatures were small with soft, jelly-like bodies. Some moved slowly, others were fixed to the sea-bed.

Sea-lilies (1) looked like strange plants. About 15 cm (6 inches) high, these animals swayed gently in the water. **Charnia** (2) looked like a bunch of feathers. **Sponges** (3) had simple bodies, open at the top. Drifting through the water were **flatworms** (4) and **jellyfish** (5). Crawling over the rocks, a **trilobite** (6) felt its way using a pair of long feelers. There were also **starfishes** (7), **worms** (8) and **lancelets** (9). Some of these creatures are unlike anything alive today.

About 400 million years ago the first plants appeared on land. Creatures such as spiders and millipedes began to live on land once the vegetation provided shelter and food.

The scene under water was very different. Some of the first sea creatures had died out, but others had developed and adapted to the changing conditions. The sea water was saltier. **Trilobites** (1) and **worms** (2) were everywhere. **Corals** (3) built great reefs. **Molluscs** (4) grazed on **algae** (5).

The major development during this period was the first appearance of a fish.

From the lancelet evolved the **jawless fish** (6). Its body armour saved this creature from being eaten by the '**sea scorpion**' (7). Jawed fish appeared several million years later. **Acanthodes** (8) was an eel-like 'spiny fish'. The giant **dunkleostus** (9) lurked on or near the sea-bed, seeking prey. None of these early forms of fish were very good at swimming.

The first fish

The **lancelet** is an odd creature that lives today in shallow seas. It is the direct descendant of the early invertebrate that evolved into fish. It moves by wriggling its very simple body. It has no head, no heart and no fins or limbs. The opening at the front end is ringed with feelers. Fossil remains of a lancelet were dated as 550 million years old.

True fish are vertebrates (animals with a backbone) that live in water. They breathe through gills. The oldest known vertebrate was **arandaspis** which lived about 500 million years ago. This fish was about 12 cm (4½ inches) long. It had no jaws or fins. The front half of its body was covered in scales.

The development of the first jawed fish was a dramatic event in animal evolution. Jaws enabled these creatures to eat larger items of food, especially other fish. **Acanthodes** swam in fresh-water rivers and lakes about 280 million years ago. It was about 30 cm (1 ft) long. Its big eyes were useful when hunting prey.

Dunkleostus, or dinichthys, was a large, fish-eating predator. There is no living creature like it. In its mouth, it had rows of bony picks and two pairs of fangs. The first of these creatures to evolve was probably only 50 cm (1⅔ ft) long. Over a period of 50 million years they developed into 9 m (30 ft) monsters.

lancelet

arandaspis

acanthodes

dunkleostus

Sharks and bony fish

About 350 million years ago, two different types of fish began to evolve. One group developed a skeleton of cartilage, instead of bone, which is lighter and softer. These creatures were the ancestors of sharks and rays. They could swim faster than other fish. In order to stay above the sea-bed they had to keep swimming, just like today's sharks. Sharks probably evolved from creatures like dunkleostus. **Cladoselache** was a prehistoric shark. This fierce killer had sharp teeth and was up to 2 m (6½ ft) in length.

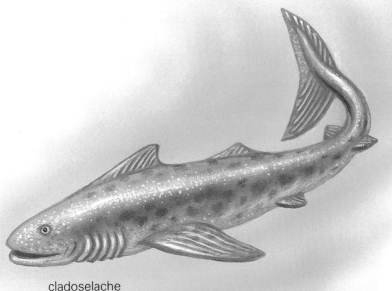

cladoselache

The second group of fish kept their bony skeleton. They evolved a new way of breathing. Part of the stomach was separated off to form a lung or swim-bladder (air pouch). Lungfish like the **dipterus** had both gills and a lung. If the rivers dried up, they could breathe in air. Some prehistoric bony fish could fill their air pouches with gas produced from their blood. By controlling the amount of gas in the pouch, the fish held its position or moved up or down. It did not have to keep moving its tail.

dipterus

The **coelacanth** first lived in oceans some 400 million years ago. It was believed to have died out until one was caught by a fisherman off the coast of Africa in 1938. This living example of a primitive fish is so like its fossil ancestors that scientists are able to learn about the bodies of early bony fish by looking at it. An adult coelacanth is 1.9 m (6¼ ft) long and feeds mainly on other fish. Fully-formed young are born from eggs that hatch inside the mother.

The invasion of the land

An important stage in the history of life took place some 350 million years ago. In a fresh-water swamp some fish began to haul themselves out of the water and on to land. The **eusthenopteron** used its front pair of fins to pull itself through the mud. Its fins had a bony skeleton (a) that did not have to undergo much change to become the limb of an amphibian. Breathing was no problem for it had air pouches like lungs. A passage linked the nostrils with the roof of the mouth. This is a feature common to all land vertebrates. The eusthenopteron was less than 60 cm (2 ft) long.

There are several reasons why lungfish might have left the water. Perhaps there was not enough oxygen in the muddy water. Perhaps they were looking for food. Perhaps their pools dried up. Whatever the reason, these creatures became more and more skilful at moving and breathing out of water. They slowly evolved into amphibians, spending part of the time in water, part of the time on land. They became the first vertebrates with legs.

An early amphibian, **ichthyostega**, was about 1 m (3 ft) long and had a tail like a fish. The fin skeleton had lost some of the bones to become more like a hand (b). It had also formed an elbow. This amphibian probably spent most of the time in water.

a

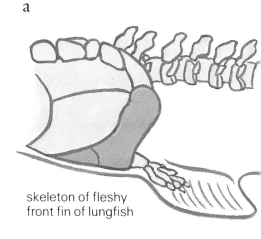

skeleton of fleshy
front fin of lungfish

b

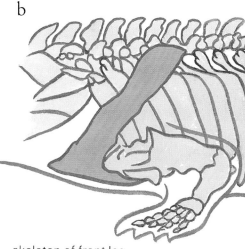

skeleton of front leg
of an early amphibian

eusthenopteron

ichthyostega

For the next 100 million years, the amphibians ruled the land. They adapted to the hot, dry conditions. Eyelids and tear glands developed to keep their eyes moist. Eventually amphibians began to produce a moisture which protected their skin and allowed them to stay out of water for longer. Some were large, like the **eryops** – 1.5 m (5 ft) in length. Its skeleton was stronger for life on land. This heavy amphibian fed on fish.

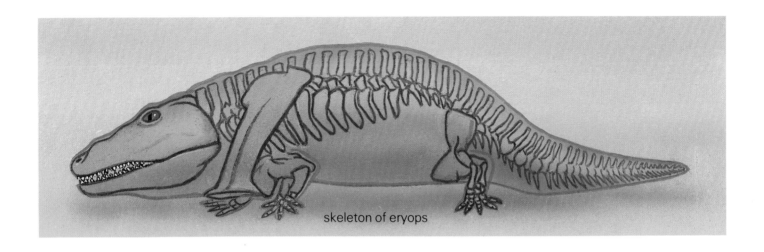
skeleton of eryops

Diplocaulus was the same size as ichthyostega. It lived in Texas, USA, and had strange 'horns' on the sides of its head. These may have helped the amphibian when swimming. They would also have made it difficult for predators to swallow it.

Mastodonsaurus was an amphibian that looked like a giant frog. It was over 3 m (10 ft) long. Its jaws were armed with sharp teeth.

Early reptiles
345–225 million years ago

The hot, wet climate resulted in the development of many plants on low-lying land. Swampy forests attracted all sorts of prehistoric animals. Amphibians evolved rapidly. Some were large, such as **eryops** (1). Others looked like newts or salamanders. Crawling in the rotting debris on the forest floor was **dolichosoma** (2). **Pantylus** (3) was an insect-eater. **Ichthyostega** (4) spent most of its time in water as did **diplocaulus** (5).

Arthropods like the **scorpion** (6) became more common. The largest ever land arthropod was a flat **millipede** (7) at 1.8 m (6 ft) long. It was during this period that the first true insects appeared. A **springtail** (8) was probably the first wingless insect. It lived in the soil and would flip into the air if scared. **Meganeura** (9) was the largest known winged insect. It had a wingspan of 70 cm (2¼ ft) and like today's dragonflies it was unable to fold its wings back. The first reptiles appeared at about 280 million years ago.

Key

1 eryops
2 dolichosoma
3 pantylus
4 ichthyostega
5 diplocaulus
6 scorpion
7 millipede
8 springtail
9 meganeura

For some 50 million years amphibians were the largest animals able to move on land. They adapted slowly to life on dry land and still had to return to water to breed. Reptiles were the first backboned animals to live entirely on land. From 280 to 225 million years ago, a number of reptiles evolved.

In the rocks of Texas, USA, there are fossils of several types of vertebrates. They are around 270 million years old. As you can see in the picture, wet-land amphibians looked like the dry-land reptiles. **Diadectes** (1) a wet-land amphibian, probably stayed close to water. It was the earliest known vertebrate to feed on plants. **Seymouria** (2) a dry-land reptile may have had a thick skin so that it could live in the desert. It only had to return to water to breed. The reptiles with huge 'sails' on their backs were **dimetrodons** (3).

The first reptiles

It is not clear which amphibians were the ancestors of reptiles. It is also not clear at which point in the history of life the developing amphibians became the first reptiles. What is known is that the most important change was the ability to lay a shelled egg. This stopped the eggs drying up after they were laid. Reptiles were now free to move away from the water. There were many more changes that helped these prehistoric animals to survive on land. Some of them are shown below.

Hylonomus was one of the first reptiles. It was 1 m (3¼ ft) long and lived in Canada. This insect-eater lived a life similar to modern lizards.

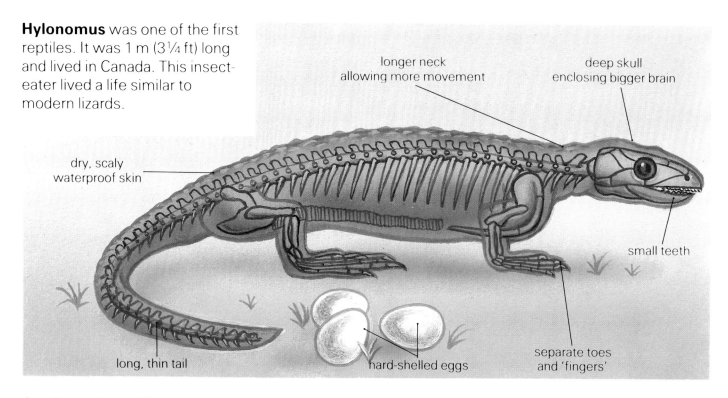

longer neck allowing more movement

deep skull enclosing bigger brain

dry, scaly waterproof skin

small teeth

long, thin tail

hard-shelled eggs

separate toes and 'fingers'

Another early reptile was **scutosaurus**. This big creature was a plant-eater. It was 2.4 m (8 ft) long and lived in Russia. Scutosaurus stood more upright than many reptiles.

A range of reptiles

Reptiles now began to invade new areas on land. In fact, they adapted to life in all parts of the world except the polar regions. Many different types developed. There were reptiles that walked, reptiles that swam, flying reptiles, running reptiles, giant reptiles and tiny reptiles, plant-eaters and meat-eaters.

Reptiles are cold-blooded animals that need heat from the sun to give them energy. **Dimetrodon** had long spines jutting from its backbone which supported a skin 'sail'. If it stood sideways to the early morning sun, many of the sun's rays hit the sail, warming the reptile's blood. This enabled it to attack prey that was still drowsy. To cool down, the dimetrodon stood with its back to the sun or in the shade, so that fewer rays hit the sail.

Proganochelys was the ancestor of the turtle. Like modern turtles, it had a heavy shell (a bony 'box' covered with horny plates). The shell was 60 cm (2 ft) long. Proganochelys was unable to pull its limbs, head or tail inside its shell. Instead these parts were protected by sharp spikes and bony knobs.

Icarosaurus was a 'flying' reptile that lived in North America. It did not really fly; it glided from tree to tree. Ribs extended from the body and were covered in skin, forming a pair of wings. The hind legs of icarosaurus were longer and stronger than its front legs. This suggests that this prehistoric lizard also ran on its two hind legs.

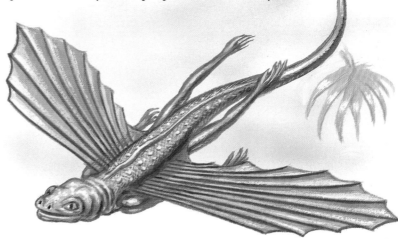

Tanystropheus was a very strange looking reptile indeed. It grew to 6 m (19½ ft) in length with a long but rather rigid neck. The young stayed on land but adults lived on the shore or in shallow waters. It probably used its long neck like a fishing rod to catch fish.

Sea reptiles

Many early reptiles adapted to a life in the open sea. The animals shown here did not actually live in the same seas or at the same time. **Ichthyosaurs** (1) were lizards that looked like modern dolphins. They had a body designed for swimming at fast speeds and they hunted in packs. Adult ichthyosaurs ate their young if there was no other food. Females gave birth to live young in the water. They could not crawl on land.

Mesosaurs (2) belonged to a family of sea lizards that seized food in sharp-toothed jaws. Fossil remains up to 8 m (26 ft) in length were uncovered in North America and New Zealand. Plesiosaurs (sea reptiles) developed into two main types. The short-necked **kronosaurus** (3) was 12 m (39 ft) long and lived near Australia. The 'snake-turtle' **elasmosaurus** (4) was up to 13 m (42½ feet) long and had over 70 neck bones. Its four flippers allowed it to make sharp turns and even swim backward but it could not dive. Elasmosaurus did not have to chase prey, it simply darted its head down into the water and grabbed passing fish. This North American reptile laid it eggs on the shore.

Reptiles rule

Lizards appeared some 230 million years ago. Apart from flying lizards and sea lizards, there were several prehistoric species living on land. Crocodiles and dinosaurs emerged from this group of large flesh-eaters. One of these, **desmatosuchus**, was 4 m (13 ft) long and had heavy skin armour. Its neck was also protected by horns.

Deinosuchus was the largest-ever crocodile at 16 m (52½ ft) long. It appeared some 140 million years ago. Fossils found in Texas show this reptile had huge jaws. It lurked in rivers and ambushed dinosaurs that came to drink. Nostrils on the top end of its snout enabled it to breathe while eating.

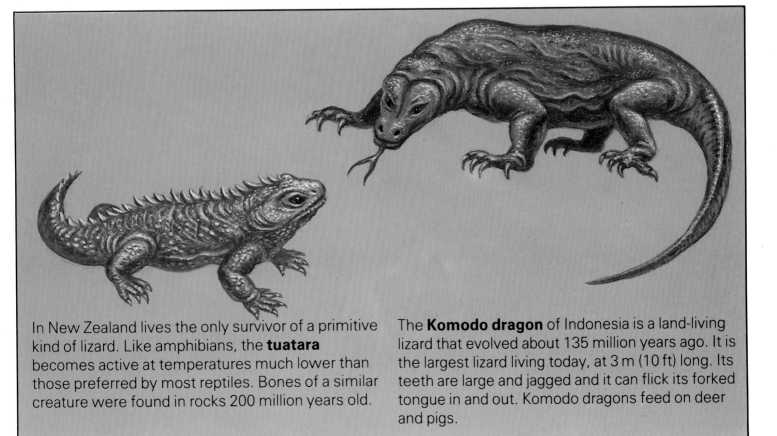

In New Zealand lives the only survivor of a primitive kind of lizard. Like amphibians, the **tuatara** becomes active at temperatures much lower than those preferred by most reptiles. Bones of a similar creature were found in rocks 200 million years old.

The **Komodo dragon** of Indonesia is a land-living lizard that evolved about 135 million years ago. It is the largest lizard living today, at 3 m (10 ft) long. Its teeth are large and jagged and it can flick its forked tongue in and out. Komodo dragons feed on deer and pigs.

Dinosaurs

225–65 million years ago

It was from 225 to 190 million years ago (the Triassic Period) that dinosaurs began to rule the land. Dinosaur is a Greek word which means 'terrible lizard'. They ruled the earth for 160 million years, finally dying out about 65 million years ago.

The climate during the Triassic Period was warm all year round. Imagine summer lasting for millions of years. Most of the land was flat. Although there were no grasses or flowering plants, the swamps and marshes were covered with a variety of trees and plants, such as yews and ferns. In the sea, **ichthyosaurs** (1) preyed on fish and other sea creatures. **Proganochelys** (2) waddled along the shore, chasing after crabs. Lakes and pools provided homes for frogs and crocodiles and the frog-like **mastodonsaurus** (3). A variety of insects crawled or flitted around. One of the first mammals, **megazostrodon** (4), was a shrew-like animal. Some early dinosaurs such as **saltoposuchus** (5) were small and some were bigger such as **plateosaurus** (6). Above their heads, reptiles like **kuehneosaurus** (7) glided from tree to tree and the first **pterosaurs** (8) flapped by on skin wings.

Earth – 225 million years ago
The land masses were joined together. Prehistoric animals could roam for thousands of miles searching for food.

Key
1 ichthyosaur
2 proganochelys
3 mastodonsaurus
4 megazostrodon
5 saltoposuchus
6 plateosaurus
7 kuehneosaurus
8 dimorphodon

During the next 60 million years (the Jurassic Period) shallow seas covered much of North America and Europe. The land masses were very slowly moving apart. Plants grew thickly, including cycads with their coloured fruits and cones.

There was an increase in the number of animals in the water, on land and in the air. By the end of the Jurassic Period, dinosaurs had become the rulers of the prehistoric animal kingdom. Some such as the **brachiosaurus** (1) had grown to enormous sizes and weights. The ground shook as they moved. The **dryosaurus** (2) and **camptosaurus** (3) fed on plants and leaves at low level. The carnivores (meat-eaters) such as **coelurus**(4) and **allosaurus** (5) were so successful that the only mammals to survive were those that came out at night or those that were small enough to burrow or climb.

In the air flapped the **rhamphorhynchus** (6) with their fragile wings of skin and bone. They probably flew slowly and were too big to fly very far. **Winged insects** (7) flew around; some have hardly changed today since that time.

Plant-eating dinosaurs

Some dinosaurs developed large hind legs and short front legs. They used their back legs for running and their front legs for grasping food.

Most dinosaurs were herbivores (plant-eaters). They had teeth shaped like pegs. When these wore down, new teeth grew in their place. **Anatosaurus** (1) had more than 1000 teeth in its wide jaws. It was a duck-billed dinosaur. Some plant-eating dinosaurs devloped into huge animals – the largest ever to live on land. **Diplodocus** (2) was the longest dinosaur. It was 26.6 m (87½ ft) long.

Plant-eating dinosaurs probably had a keen sense of smell. This helped them to know when predators were near. Some ran away. Others, like diplodocus, would lash

out with their tails. If that failed to scare off the attacker, the dinosaur would rear up and crash down its front legs (as an elephant does to crush a tiger). Other dinosaurs had horny plates and spikes. **Triceratops** (3) had three horns. At 9 m (30 ft) from beak to tip of tail, it was one of the largest horned dinosaurs. When a triceratops put its head down and charged, predators probably ran away.

Herds of bonehead dinosaurs also roamed the plains eating the vegetation. Two adult male **stegoceras** (4) charged towards each other and banged their heads together. The winner of the duel probably ruled the herd of females. These boneheads had a brain the size of a hen's eggs inside a bony dome five times thicker than the skull of a human.

Meat-eating dinosaurs

Most carnivores in the dinosaur world walked and ran on their hind legs. Many of these speedy runners had a stiffened tail. This helped them to balance on their strong hind legs. Some carnivores were no bigger than a chicken. Only the larger meat-eaters were a real threat to the plant-eating dinosaurs.

The dinosaur chasing a lizard is a **compsognathus** (1). We know this carnivore caught such prey because its fossilised rib-cage has been found containing lizard bones. Smaller than a chicken, it was 60 cm (2 ft) long and weighed only 4 kg (6½ lb). Compsognathus was possibly the ancestor of birds (see page 31).

Tyrannosaurus rex (2) was king of the carnivores. It was 12 m (39 ft) from nose to tail. It was probably not a fast runner. Tyrannosaurus tended to feed on dead or injured dinosaurs. If it caught live prey, its teeth might have snapped off during the fight. The sharp claws on its hind feet were

used as carving knives and it tore off lumps of flesh with its fangs. Strangely, its tiny arms were too weak to use as weapons and too short to lift food to its mouth.

The name **deinonychus** (3) means 'terrible claw'. This ferocious carnivore had the first toe turned back and the second toe had a claw 12.5 cm (5 inches) long. When it caught up with its prey, deinonychus would leap up, lashing out with its claws and pin it to the ground.

This meat-eater was 3 m (10 ft) long.

The 8 m (26 ft) long herbivore, **stegosaurus** (4) would have been preyed on by the 10 m (32½ft) long **allosaurus** (5). Its plates made the stegosaurus look bigger than it actually was and the allosaurus was nimble enough to dodge the spikes on the stegosaurus' tail. The allosaurus had razor-sharp teeth and large claws. Its huge jaws could gape wide enough for it to shove great chunks of meat inside.

Dinosaurs rule

Many modern plants and animals appeared for the first time during the last period of the Dinosaurs, called the Cretaceous Period, from 135 to 65 million years ago. Flowering plants spread to many parts and oak trees grew near streams. The weather was generally warmer than it is now, as climates ranged from very hot at the Equator to warm in the North.

Dinosaurs reached a peak as more and more plant-eaters evolved with teeth designed for chewing the new tough leaves. Each creature had its own special feeding ground. The horned dinosaurs like the **triceratops** (1) kept to higher land. The **anatosaurus** (2) and other duckbills ate so much they cleared some areas completely. Large carnivores such as **tyrannosaurus** (3) attacked and ate herbivorous dinosaurs like duckbills.

By 70 million years ago, **snakes** (4) had evolved from lizards. The most obvious differences was that these reptiles had lost their front and rear limbs. This may be because they began to burrow underground, feeding on worms and insects. The big marine reptiles, **mesosaurs** (5) were flippered lizards. Overhead flew **pterosaurs** (6) plus many types of **true birds** (7 and 8).

Key							
1	triceratops	3	tyrannosaurus rex	5	mesosaur	7	ichthyornis
2	anatosaurus	4	burrowing snake	6	pterosaur	8	wader birds

Gliding animals

The first creatures to fly were insects. In order to catch this food supply, prehistoric animals had to adapt to become skilful fliers themselves. Other land animals learnt to fly as a means of escape from their flesh-eating enemies. The first vertebrates to take to the air may have been descended from a tree-dweller known as a **podopteryx**. A web of skin linked the tail, hind legs and

forelimbs. Podopteryx used this skin 'parachute' to glide, like today's flying lizards and flying squirrels.

Pterosaurs were winged lizards. Their wings developed on their front limbs. The skin was attached to the long bones of the fourth finger on each hand and it curved back to the hind legs. Their muscles were too weak to flap their wings, so pterosaurs glided on warm air currents. They fed on fish and small lizards. Pterosaurs would have shuffled along on land because their legs were very short. The early pterosaurs had long wings and tails.
Rhamphorhynchus (1) was the size of an eagle. It probably used its long tail for steering when flying.

The later pterosaurs had long wings, little or no tail and a toothless beak. A fully grown **pteranodon** (2) had a wing span of 7.6m (25ft).

Quetzalcoatlus was the largest creature ever to have flown. Its wingspan was 12m (39ft) from wingtip to wingtip. No one knows how a beast that big got up into the air. It might have hauled itself to the top of a tree or cliff before diving off.

New Life

65 million–5,000 years ago

All dinosaurs and most reptiles vanished from the face of the Earth around 65 million years ago. Why did this happen? Perhaps there were natural disasters such as floods or earthquakes. Perhaps carnivores ate all the herbivores and then ate each other. It was more probably caused by a change in the climate that made it too cold for the dinosaurs and they could not adapt. Whatever the cause, the death of the dinosaurs signalled the start of the New Life, the Age of Mammals.

During this period, mammals evolved shapes and sizes that suited them for life in almost all places. Some took to the air (1), others went into water. The first hoofed herbivores (2) and the first rodents (3) appeared. As grasses grew, so grazing mammals evolved – beasts that looked like horses (4), camels (5) and rhinos (6).

Large flightless birds (7) began to evolve in the places where there were no big reptiles or mammals that could prey on them. Some birds (8) took to hunting prey at night. **Ichthyornis** (9) was the first known bird with a breastbone designed to support the muscles needed for flying. The **long-legged wader** (10) was the ancestor of ducks and geese.

Earth – 40 million years ago
As land masses moved apart, climates cooled. Different types of mammals developed in different lands.

Key

1 bats
2 uintatherium
3 paramys
4 hyracotherium
5 alticamelus
6 arsinoitherium
7 diatryma
8 ogygoptynx
9 ichthyornis
10 proardea

Birds

The prehistoric animal which links birds to their dinosaur ancestors was probably **archaeopteryx**. It had solid bones like a reptile, yet it had feathered wings like a bird. Three clawed fingers sprouted from each 'feathered arm'. Archaeopteryx was too big to fly and it was not a good glider as its wings were too short. It was probably a fast runner, able to catch flying insects.

Hesperornis, an early true bird, had very small wings. It was clumsy on land, using its tail to help it balance. Its webbed feet were used for swimming. Hesperornis seized fish in its toothed jaws.

A large flightless bird stalked the plains of North America 50 million years ago. Taller than an adult human, **diatryma** had a huge beak which made it look fierce. Its body and small wings were covered in fine feathers.

archaeopteryx

Although these birds died out millions of years ago, there are still large flightless birds alive today. The most famous are the **ostrich**, emu and cassowary.

hesperornis

diatryma

31

Mammals

Not all reptiles could adapt to the new colder conditions. However, mammals had evolved with the ability to keep their bodies warm at all times. This was helped by the body covering of hair. Being warm-blooded allowed some mammals to carry on hunting prey at night. They also looked after their young.

Only fossil teeth and jaws have been found of these early mammals. The first mammals were insect-eating shrews, like **megazostrodon**.

We know about early mammals from looking at the primitive mammals that are alive today. The **echidna** is a spiny anteater. It has no ears and has little control over its body temperature. A female echidna places leathery eggs at the bottom of a closed burrow and keeps them warm for two weeks. The tiny young then attach themselves to hairs in the mother's pouch.

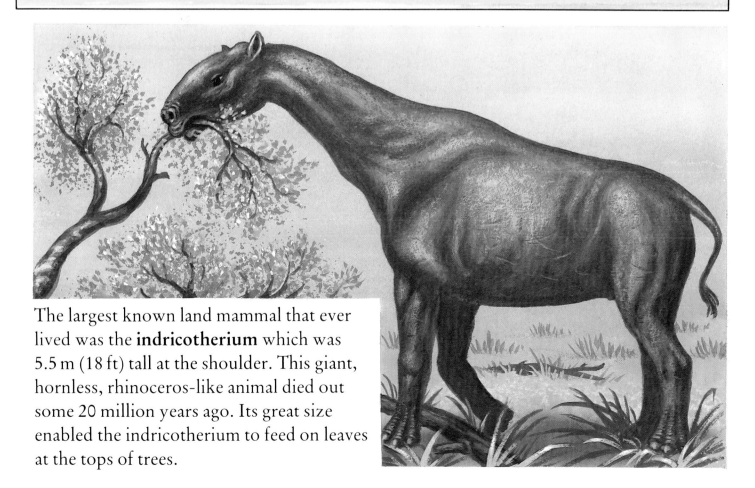

The largest known land mammal that ever lived was the **indricotherium** which was 5.5 m (18 ft) tall at the shoulder. This giant, hornless, rhinoceros-like animal died out some 20 million years ago. Its great size enabled the indricotherium to feed on leaves at the tops of trees.

The most successful mammals in terms of numbers and variety are the rodents. This group includes rats, squirrels, beavers, porcupines and many more. Rodents live in trees, on mountains, under the ground and in streams. The first known rodent was **paramys** (1). It was a climber like a squirrel.

Possibly the largest flesh-eating mammal ever was **megistotherium** (2). Its head was twice as big as any bear's. This monster weighed about 900 kg (1980 lb).

Uintatherium was a hoofed animal and stood 2 m (6½ ft) at the shoulder. This slow-moving plant-eater lived in North America. It had three pairs of horny swellings on top of its skull. Males also had a pair of tusks that rested on another bony outgrowth.

Mammals rule

By 35 million years ago large flesh-eating mammals ruled most continents. They had bigger brains, keener ears and were faster than other mammals. In the picture below, a very big cat, **smilodon** (1) is on top of a large prehistoric herbivore, **megatherium** (2). Smilodon was an American 'sabre-tooth tiger'. It used strong neck muscles to stab its teeth into prey. Megatherium was a huge ground sloth, 6 m (20 ft) long and nearly twice as tall as an adult human. It walked on its knuckles and the sides of its feet. When it reared up, its strong tail acted as a prop. Then it could claw leaves into its mouth. Megatherium probably died out after humans invaded the forests where they lived. It is possible that sabre-tooth tigers caught and killed creatures as large as the **mastodon** (3). This prehistoric elephant had a furry coat of red hair.

Around 20 million years ago there were many animals in North America. Those shown here all have hoofed feet and chew their food. **Alticamelus** (1) was a prehistoric camel-giraffe. If it lifted its head it could feed on leafy twigs 3 m (10 ft) off the ground. **Synthetoceras** (2) was a prehistoric type of deer. Males grew Y-shaped horns jutting up from the nose and a pair of shorter horns behind the eyes. **Brontotherium** (3) was between a rhinoceros and an elephant in size. It had three hoofed toes on each hind foot and four toes on each front foot. Males probably used their horns when fighting.

More mammals

The first known horse was about the size of a fox. **Hyracotherium** or 'dawn horse' was only 40 cm (1¼ ft) tall at the shoulder. It munched soft leaves 50 million years ago in swampy North American and European forests before the continents finally separated. Its descendants died out in Europe but continued to evolve in North America. Later stages crossed back into Europe and continued their evolution, but the horse mysteriously died out in America. The modern horse, **equus**, evolved about 2 million years ago. It spread into most continents but was reintroduced into America only a few hundred years ago by man. Wild species alive today include Przewalski's horse of Mongolia, the wild ass and the zebra.

Certain mammals returned to a life in water. **Basilosaurus** (1) was an early whale – up to 20 m (66 ft) long. It had saw-edged teeth. Its front limbs were flippers and there were no hind limbs. The early relative of sea-lions was **allodesmus** (2). It looked like the largest seal alive today, the sea-elephant.

The **cave bear** was a large member of the dog family. At about 4 m (13 ft) tall on its hind feet, it was well over a third larger than even the biggest of the modern brown bears. Rounded teeth indicate that the cave bear fed mainly on plants. Its home was often in the mountains of Europe.

The **mammoth** (1) and the **woolly rhinoceros** (2) lived on the icy tundra within sight of the great ice sheets during the ice ages in Europe and North America. They both had long hair and thick underfur. The woolly mammoth used its tusks to defend itself and possibly to clear snow off plant food. The woolly rhino was able to dig up roots with its large horn. There is evidence that mammoths also lived in other continents. Cave paintings show mammoths being hunted by early cave-people. Mammoths died out about 10,000 years ago. The modern Indian elephant is the closest living relative to the mammoths.

The arrival of mankind

About 65 million years ago, a group of shrew-like animals began to live in the trees. This may have been to escape predators. These animals evolved into the first example of a primate 20 million years ago, called **prosimian** (1) – the ancestor of humans as well as apes and monkeys.

From prosimian came two branches, one was **ramapithecus** (2), the ancestor of mankind and the other was dryopithecus, the ancestor of monkeys and chimpanzees. The chimpanzee is therefore our 'cousin' and probably our closest relative in the animal kingdom.

Ramapithecus looked more like an ape than a man and lived about 10 million years ago. Its skeleton structure shows that after several million years it was walking upright

and therefore probably using its hands to pick roots and fruits or to hold a stick or stone as a weapon.

Between 4 million and 1½ million years ago, **australopithecus** (3) evolved. It was definitely more human-like. It lived in groups and was starting to eat meat as well as vegetation. It therefore had to find better ways to kill. It was not big, up to 1.2 m (4 ft) tall, nor very strong, so it had to invent weapons and learn to hunt small animals. It probably used old bones as clubs.

About 2 million years ago, the first human being appeared. His skeleton was very close to ours. He was called **homo habilis** (handy man) (4). He made stone weapons and tools. He began to make very primitive shelters. He was about 1.3 m

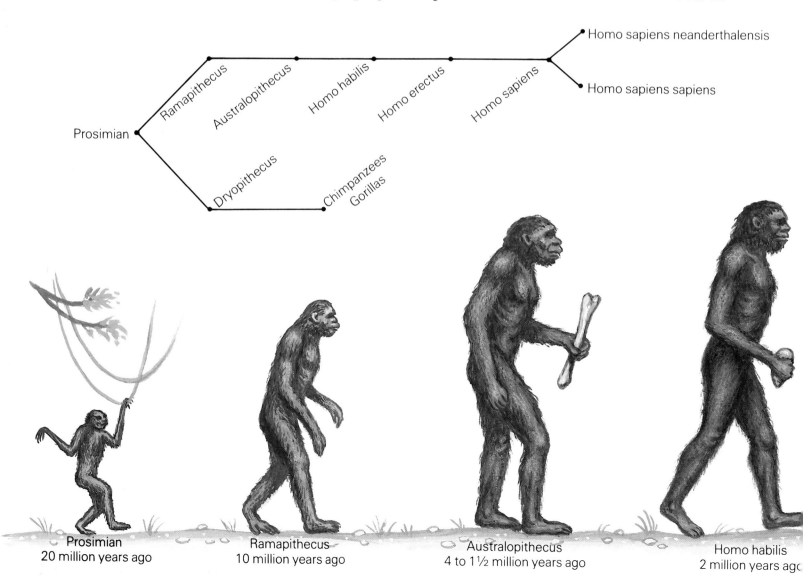

Prosimian
20 million years ago

Ramapithecus
10 million years ago

Australopithecus
4 to 1½ million years ago

Homo habilis
2 million years ago

(4¼ ft) tall. His successor was **homo erectus** (upright man) (5), who appeared between 1 million and 800,000 years ago. He was continually improving stone tools, mainly by putting handles on them. He was taller and stronger than homo habilis. His body and limbs were very similar to ours, but the skull was still changing. As he became more intelligent, homo erectus used his hands more and made tools to do what his jaws had done before. Over a long period, his jaws became smaller making him look more like a modern man. His greatest discovery was fire about 400,000 years ago. In China, a fireplace was found with a pile of ashes that were 6 m (19½ ft) deep, showing that the fire had been kept continually burning for many generations.

Homo erectus gradually evolved into **homo sapiens** (wise man) (6) and for a long time groups of both types of primitive human beings lived alongside each other until homo erectus died out leaving homo sapiens. One form of homo sapiens, **Neanderthal man** (7), appeared about 150,000 to 100,000 years ago and died out about 35,000 years ago. He was about 1.6 m (5 ft) tall.

About 45,000 years ago, the last link appeared. This was **homo sapiens sapiens** (8) who was probably responsible for the extinction of Neanderthal man. He was physically no different from humans today. From then on as his intelligence increased, he travelled further, built homes and eventually, 5,000 years ago, began recording his activities in ancient writing.

Homo erectus
1 million years ago

Homo sapiens
800,000 years ago

Homo sapiens neanderthalensis
150,000 years ago

Homo sapiens sapiens
45,000 years ago

Neanderthal man

The best-known primitive human is Neanderthal man. He lived from about 150,000 to 35,000 years ago.

He was similar in build to us but a little shorter and stockier. His face was broad, with a heavy brow over his eyes and no chin. He probably had a very basic language. He lived in a group probably made up of several families. Each winter they would return to a permanent winter home after following the herds of mammoth and woolly rhinoceroses all summer. Neanderthal man was a skilful hunter, using spears and clubs to bring down these big animals. He wore skins to keep warm in the cold conditions near the ice sheets where he lived during the Ice Age.

Fire was used for protection as well as for warmth. A fire in an entrance to a cave would deter most animals. They also used it for cooking and for hardening wooden spears.

They buried their dead, often with tools and flowers. Sometimes bones of an ibex, cave bear or lion were buried with them, suggesting that these animals may have been sacred to Neanderthal man.

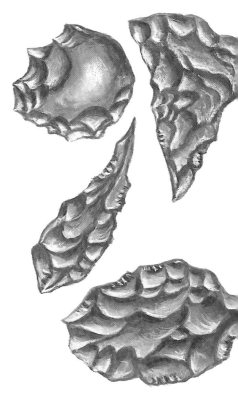

Primitive people were quite skilled at making stone tools and weapons of various sizes. Hand-axes, scrapers, knives, spikes and drills were chipped out of stone.

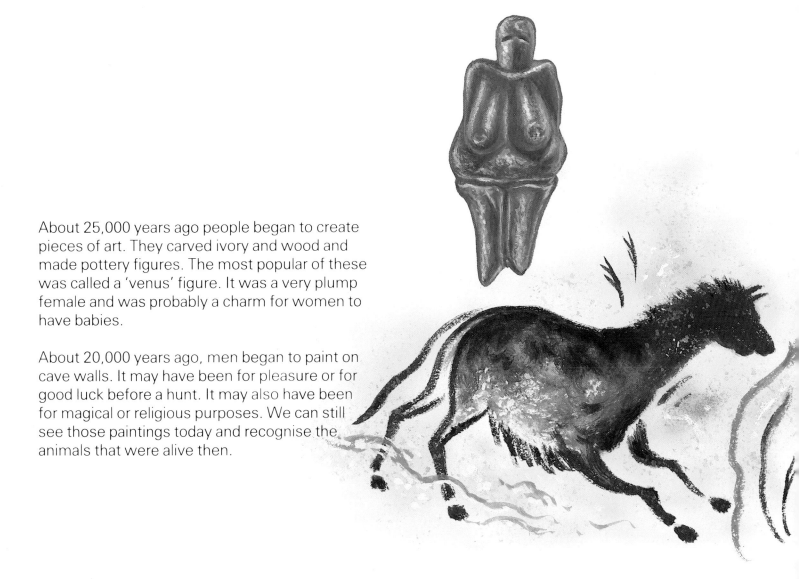

About 25,000 years ago people began to create pieces of art. They carved ivory and wood and made pottery figures. The most popular of these was called a 'venus' figure. It was a very plump female and was probably a charm for women to have babies.

About 20,000 years ago, men began to paint on cave walls. It may have been for pleasure or for good luck before a hunt. It may also have been for magical or religious purposes. We can still see those paintings today and recognise the animals that were alive then.

Fossils

When animals die their bodies rot, often leaving no trace. But, if conditions are right, some parts may be preserved. An animal may have died and been buried in a seabed, a riverbed, a desert or even under snow. Layers of mud, sand or ice would have built up over the years burying the remains deep below ground. The hard parts of the animals may be preserved in the rock as fossils; some animals could even be preserved whole in ice. Slow wearing away of rock or sudden earth movements such as earthquakes can expose the deep layers and reveal those fossils that have been hidden for thousands or millions of years. By studying these remains, scientists are able to trace the history of life on earth and the links between the ancient and modern animal species.

Traces of the earliest forms of life have been found in Australia. Fossilised layers of limestone created by **blue-green algae** date back 3,500 million years. Embedded in sandstone, also in Australia, the remains of 600 million year old **jellyfish** have been found.

Fish fossils have been found dating back as far as about 430 million years ago. An amazing find at the site of an ancient lake in Italy revealed over 100,000 fossilised **fish**; the earliest of them were dated as 55 million years old. They lived in a lake overshadowed by a volcano. Each time the volcano erupted, the fish in the lake were 'cooked' and their remains buried. This left several layers of fossilised fish.

Insects are extremely frail and therefore do not often preserve as fossils, except under the most ideal conditions (being covered with a layer of mud that dries hard). Some insect finds on land have been traced back to about 345 million years ago, but the best finds are of insects trapped in tree resin (sap) which has hardened to form amber. The best fossil **insects in amber** were found by the Baltic Sea and are about 30 million years old.

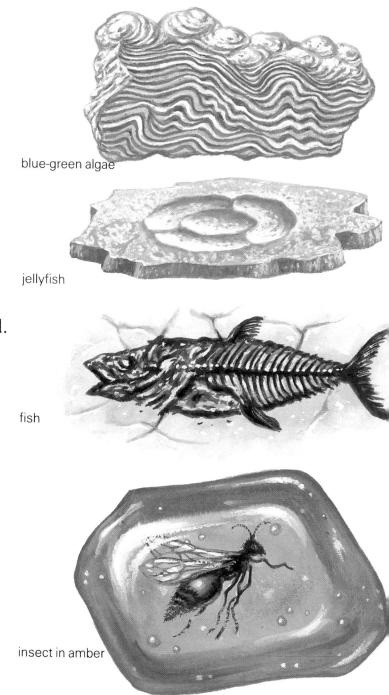

blue-green algae

jellyfish

fish

insect in amber

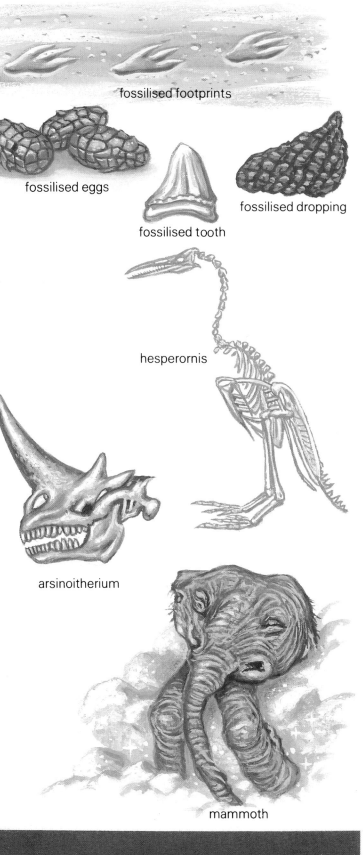

fossilised footprints

fossilised eggs

fossilised tooth

fossilised dropping

hesperornis

arsinoitherium

mammoth

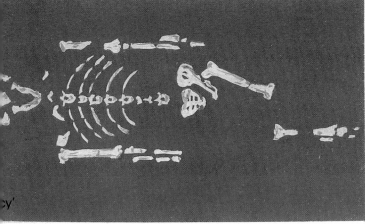

The age of the reptiles, particularly that of the dinosaurs, has been a rich source of fossil finds. From these, scientists can build up a full picture of what the dinosaurs were like. The spacing of **fossilised footprints** tells them how fast they moved; bones tell them how big the dinosaurs were. **Fossilised eggs**, some with babies still inside, supply information about their breeding habits; **fossilised teeth** and **droppings** give clues to what the dinosaurs ate. The skeletons of 31 iguanodon found together in a Belgian coal mine support the view that some herbivores probably roamed in herds.

Like insects, bird fossils are rare because of the frailty of the bones, but the earliest true birds have been dated at about 80 million years old. One was **hesperornis**, for which all the bones have been found and pieced together.

Fossils from the age of mammals have provided evidence of animals which are now extinct and also of the ancestors of some modern day animals. The double-horned **arsinoitherium**, which has no living descendant, has been found in Egypt and dated as 37 million years old. **Mammoths** and woolly rhinoceroses, ancestors of the Indian elephant and rhinoceros, have been found deep-frozen in Siberia.

Evidence of our own primitive ancestors has also been found. This can be traced back about 60 million years to a shrew-like tree-living primate. **'Lucy'** (left) found in Africa, is the most complete australopithecus skeleton ever found. This has been dated as 3½ million years old. It clearly shows the link between humans and our ape-like ancestors.

Index

The creatures in this index have a short description and Latin names are translated in *italic*.

This diagram shows when
different creatures evolved
and the important Ages of
Life on Earth.

Blue gree

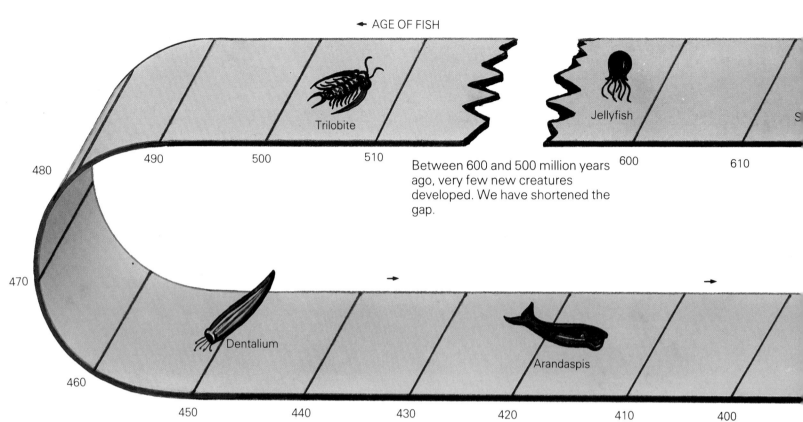

← AGE OF FISH

Trilobite

Jellyfish

S

490   500   510

480

470

460

450   440   430   420   410   400

Between 600 and 500 million years
ago, very few new creatures
developed. We have shortened the
gap.

600   610

Dentalium

Arandaspis

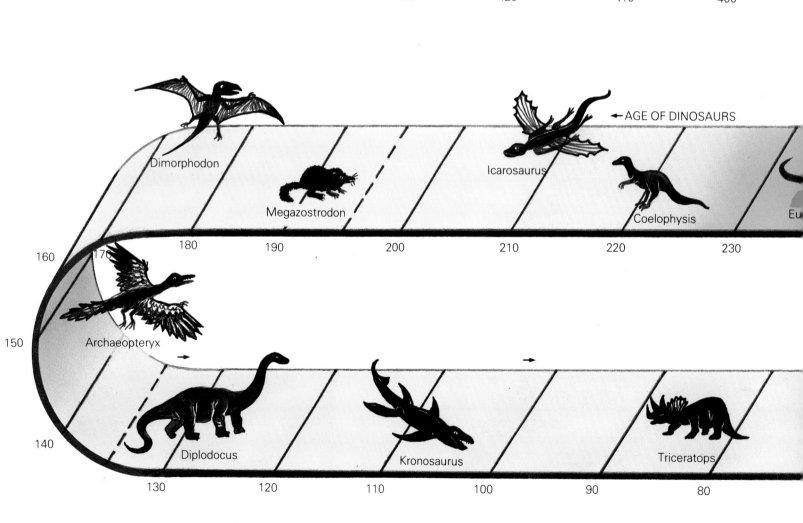

Dimorphodon

Megazostrodon

Icarosaurus

← AGE OF DINOSAURS

Coelophysis

Eu

180   190   200   210   220   230

160   170

150

140

Archaeopteryx

Diplodocus

Kronosaurus

Triceratops

130   120   110   100   90   80